THE ULTIMATE

Ocean

FACTS

for Kids, Teens, & Adults

**Dive Into Underwater Wonders,
Discover Mind-Blowing Creatures,
Phenomena, Records, Inventions,
and the Mysteries of the Deep!**

**Book 5 of Eleven Worlds to Explore
Ethereal Ray**

Copyright © 2024 Ethereal Ray
All rights reserved.

No part of this publication may be reproduced, stored in a retrieval system, or transmitted in any form or by any means, electronic, mechanical, photocopying, recording, or otherwise, without the prior written permission of the author, except for brief quotations in reviews or scholarly analysis.

Marine animals designed by Freepik

Marine animals designed by brgfx on Freepik
Coral & Finn image by catalyststuff on Freepik

Table of Contents

Chapter 1: Welcome to the Ocean! 1

Chapter 2: Exploring the Ocean's Layers and Regions...... 6

Chapter 3: Marine Life - Creatures Big and Small.......... 20

Chapter 4: Coral Reefs - The Rainforests of the Sea 29

Chapter 5: Ocean Exploration - Boldly Diving Deep 38

Chapter 6: Incredible Ocean Inventions 44

Chapter 7: Ocean Phenomena and Mysteries 52

Chapter 8: Ocean Careers - Your Path to Marine Adventures 59

Chapter 9: Legends of the Sea - Mythology and Folklore 66

Chapter 10: Beyond the Blue - Ocean Mysteries and Phenomena 78

Ocean Quiz 85

Glossary of Ocean Terms... 93

Review Request 99

Hi there! I'm **Coral**, and this is my **sharp-toothed but friendly shark buddy, Finn!** We're your guides on this **exciting underwater adventure**. Together, we'll explore the **mysteries of the deep**, meet **fascinating sea creatures**, and uncover **hidden treasures beneath the waves!**

If you enjoyed this **journey into the ocean**, maybe you could ask a grown-up to **leave a review on Amazon?** Reviews help us **dive even deeper** and create even more **amazing ocean books** for curious readers like you!

Scan the QR below!

"The sea, once it casts its spell, holds one in its net of wonder forever

- Jacques Cousteau

Get ready to dive into an incredible journey through the wonders of the ocean! **"The Ultimate Ocean Facts for Kids, Teens, & Adults: Dive Into Underwater Wonders, Discover Mind-Blowing Creatures, Phenomena, Records, Inventions, and the Mysteries of the Deep!"** is packed with fascinating information, stunning visuals, and engaging activities designed to ignite your curiosity and deepen your understanding of the underwater world.

This book is more than just a collection of facts—it's your passport to the ocean:

- **Ignite your imagination**: Explore vibrant coral reefs, meet the ocean's giants like whales and sharks, and uncover the mysteries of the deep-sea trenches. Each chapter will spark your curiosity and transport you to the magical world beneath the waves.

- **Fuel your mind**: Learn how marine creatures adapt to their environments, discover incredible underwater inventions, and delve into the fascinating science of ocean currents and ecosystems. This book is a knowledge treasure chest for curious minds of all ages.
- **Expand your horizons**: From the tiniest plankton to the largest blue whale, uncover the secrets of the ocean and gain a deeper appreciation for the life and phenomena beneath the surface.
- **Bond through discovery**: Share this ocean adventure with family and friends. Explore the wonders of the underwater world together, create your own marine life crafts, and embark on a journey of learning and discovery.

Embark on an exploration of the ocean's incredible biodiversity, stunning landscapes, and fascinating phenomena. With every page you turn, you'll uncover new wonders and deepen your connection to this amazing underwater realm.

So, grab your snorkel, put on your flippers, and get ready to dive in! Your aquatic adventure awaits!

Chapter 1:
Welcome to the Ocean!

Have you ever stood by the shore and wondered, "What mysteries lie beneath the waves?"

In this chapter, we'll uncover mind-blowing facts about the ocean, from its vast size to the incredible creatures that call it home.

We'll discover why the ocean is so important to life on Earth and learn about its hidden wonders, from colorful coral reefs to the deepest, darkest trenches.

So, grab your snorkels, fire up your imaginations, and prepare to make a splash! It's time to embark on an aquatic adventure like no other!

What is the ocean?

The ocean is a vast and mysterious world that covers more than 70% of our planet Earth. It's where millions of marine creatures live, where waves dance with the shore, and where the mysteries of the deep remain unexplored. The ocean isn't just one big body of water—it has five main parts: the **Pacific**, **Atlantic**, **Indian**, **Southern**, and **Arctic Oceans.** They work together to support life on Earth in amazing ways!

Why Is the Ocean Important?

The ocean isn't just beautiful—it's vital for life on Earth.

- **Oxygen Factory:** Over 50% of the oxygen we breathe comes from tiny ocean plants called phytoplankton. Without them, we wouldn't survive!

- **Climate Control:** The ocean absorbs heat from the Sun and redistributes it through currents, which helps regulate the Earth's temperature.

- **A Source of Food and Medicine:** Millions of people depend on the ocean for seafood, and scientists are discovering life-saving medicines from marine organisms.

In the ocean, you'll find amazing things like:

- **Coral Reefs:** Vibrant underwater cities filled with colorful fish, sea turtles, and tiny creatures called polyps that build reefs.

- **Marine Giants:** From the majestic blue whale, the largest animal on Earth, to the gentle giants like manta rays and whale sharks.

- **Tiny Wonders:** Microscopic plankton, which not only provide food for many ocean creatures but also produce most of the oxygen we breathe!

- **Hidden Trenches:** Mysterious underwater valleys like the Mariana Trench, the deepest part of the ocean, where bizarre creatures live in total darkness.

- **Underwater Volcanoes:** Known as seamounts, these volcanoes erupt beneath the waves, creating new islands over time.

But wait, there's more! The ocean is also home to fascinating phenomena:

- **Bioluminescence:** Creatures like jellyfish and plankton that glow in the dark, lighting up the water with an eerie blue-green glow.

- **Tides and Currents:** The ocean's "heartbeat," controlled by the Moon's gravity and the Earth's rotation, which shapes our coasts and even the weather.

- **Ocean Sounds:** From the haunting songs of humpback whales to the clicks and whistles of dolphins, the ocean is alive with sound.

How Big Is the Ocean?

The ocean is so mind-bogglingly vast that it's hard to wrap our heads around! Here are some fun ways to imagine its size:

- If you poured all the water in the ocean into a single glass, that glass would hold 97% of all the water on Earth!

- The Pacific Ocean alone is bigger than all the continents on Earth combined.

- The average depth of the ocean is about 12,100 feet, but some trenches plunge more than 36,000 feet—deeper than Mount Everest is tall!

Despite its size, we've explored less than 5% of the ocean, leaving the vast majority a complete mystery. Who knows what other creatures, landscapes, or treasures are waiting to be discovered?

What Lies Beneath?

The ocean is divided into five main zones, **each with its own unique characteristics and marine life:**

1. **Sunlight Zone:** The top layer, where sunlight penetrates and most marine life thrives. Coral reefs, turtles, and schools of fish live here.

2. **Twilight Zone:** Below the sunlight, this dimly lit layer is home to creatures like lanternfish and giant squids.

3. **Midnight Zone:** Pitch-black and cold, this zone is where bioluminescent creatures light up the darkness.

4. **The Abyss:** The ocean floor, with extreme pressure and fascinating creatures like sea cucumbers and deep-sea shrimp.

5. **The Trenches:** The deepest parts of the ocean, like the Mariana Trench, where life adapts to crushing pressure and total darkness.

Fun Facts About the Ocean

- The ocean is home to 94% of all life on Earth, yet we've only discovered a fraction of it.

- Dolphins have names for each other! They use unique whistles to call out to their friends.

- There are underwater rivers and waterfalls—caused by differences in water temperature and salinity.

An Ocean Perspective

The ocean reminds us of the beauty and fragility of our planet. It connects continents, supports ecosystems, and inspires a sense of wonder.

- One Global Ocean: Just like astronauts see Earth as one interconnected world, the ocean connects us all, with no boundaries.

- Protecting Our Oceans: The more we learn about the ocean, the more we understand the importance of conserving it for future generations.

Are you ready to dive into this amazing underwater world? Let's explore the ocean's wonders, discover its secrets, and meet the incredible creatures that call it home. Our journey begins now—grab your snorkels and let's dive in!

Chapter 2:
Exploring the Ocean's Layers and Regions

When you think of the ocean, you might imagine waves gently lapping at the shore or colorful fish swimming through coral reefs.

But did you know that the ocean is so deep and vast that we've explored less than **5%** of it? That means most of it remains a mystery!

The ocean is more than just water—it's a world of layers and regions, each with its own unique environment and inhabitants. From the sunlit surface to the darkest depths, the ocean teems with life, natural phenomena, and incredible stories waiting to be uncovered.

Let's dive deeper into this vast, blue universe!

The Ocean Zones: Layers of Mystery and Life

The ocean is a vast and mysterious place, with each layer or zone offering a unique environment for marine life. From the sun-drenched surface to the crushing depths, each zone is home to specialized creatures and ecosystems, many of which have adapted to survive in extreme conditions. Let's dive deeper into these zones and explore the wonders they hold!

1. The Sunlight Zone (Epipelagic Zone)

- **Depth:** Surface to 200 meters (656 feet).

- **Characteristics:** This is the top layer of the ocean, where sunlight can penetrate the water and create a vibrant, colorful world. The sunlight zone is the most hospitable area for marine life because of the warmth and light that sustain life. Here, you'll find coral reefs, kelp forests, and thriving ecosystems where the majority of ocean species live.

- **Marine Life:**
    - **Coral Reefs:** Coral reefs are underwater "cities" made up of tiny creatures called polyps. They provide shelter and food for over 25% of all marine species, including clownfish, parrotfish, and moray eels.
    - **Sea Turtles:** These ancient creatures are wanderers of the ocean, migrating thousands of miles from nesting beaches to feeding grounds.
    - **Dolphins and Whales:** Dolphins are social creatures, often traveling in pods, while humpback whales are known for their songs and long migrations across oceans.

- **Jellyfish:** These strange creatures float through the water, and while some are beautiful and harmless, others can deliver powerful stings.

Fun Fact: The sunlight zone is sometimes called the "**Euphotic Zone**" because it's the only layer where photosynthesis can occur, as sunlight penetrates deep enough to support plant life like phytoplankton, which is essential for life on Earth. These microscopic plants produce **over 50% of the oxygen** we breathe!

The Twilight Zone (Mesopelagic Zone)

- **Depth: 200 meters to 1,000 meters** (656 feet to 3,280 feet).

- **Characteristics:** This is the zone where light begins to fade. It's not completely dark, but sunlight is weak and not sufficient for photosynthesis. The twilight zone is a mysterious realm, and the creatures that live here have evolved to thrive in the dim light, adapting to the lower visibility and colder temperatures. Here, life is more spread out, and many animals have developed unique features, such as bioluminescence (the ability to make their own light).

- **Marine Life:**
  - **Lanternfish:** These small, silver fish get their name from the glowing organs they use to communicate and attract prey.
  - **Squid:** Species like the **vampire squid** and **bioluminescent squid** are found in this zone. Some use their glow to confuse predators or attract mates.
  - **Anglerfish:** These predators are most famous for the glowing lure that dangles from their head, attracting unsuspecting prey.

- **Sperm Whales:** These giant mammals are one of the few creatures able to dive deep into the twilight zone in search of food like giant squid.

Fun Fact: Some animals in the twilight zone have specialized eyes to help them see in low light, and creatures like **bioluminescent jellyfish** light up the water, creating a mesmerizing effect.

3. The Midnight Zone (Bathypelagic Zone)

- **Depth: 1,000 meters to 4,000 meters** (3,280 feet to 13,123 feet).

- **Characteristics:** This zone is pitch black. No sunlight reaches here, and it's incredibly cold with water temperatures near freezing. Despite the lack of light and harsh conditions, the midnight zone is home to some of the most unusual and unique creatures on Earth. Inhabitants of this zone have adapted to the extreme cold and crushing pressure, developing specialized bodies and behaviors.

- **Marine Life:**
    - **Giant Squid:** These mysterious creatures have fascinated humans for centuries, with tentacles that can reach up to **40 feet** in length. They are believed to be some of the largest invertebrates in the world.
    - **Anglerfish:** Not just found in the twilight zone, some anglerfish live in the midnight zone, where they use their glowing lures to attract prey in the complete darkness.
    - **Viperfish:** With long, needle-like teeth and large eyes that help them catch prey in the

dark, viperfish are true hunters of the midnight zone.

- o **Goblin Sharks**: These eerie sharks are often referred to as "living fossils" and are rarely seen by humans. They have elongated snouts that help them sense prey in the deep.

Fun Fact: The Midnight Zone has been described as the "last frontier" of the ocean, a place where new species are constantly being discovered, and its secrets are still largely unknown.

4. The Abyss (Abyssopelagic Zone)

- **Depth: 4,000 meters to 6,000 meters** (13,123 feet to 19,685 feet).

- **Characteristics**: The abyss is a cold, dark place with no sunlight whatsoever. The water pressure is over **600 times greater** than at the surface, and temperatures hover just above freezing. Despite these extreme conditions, life flourishes here. The abyss is home to some of the most resilient creatures on Earth, creatures that have adapted to the near-absolute darkness and crushing pressure.

- **Marine Life**:

    - o **Sea Cucumbers**: These soft, sausage-shaped creatures are scavengers, playing a critical role in cleaning the ocean floor by eating dead organisms.

    - o **Deep-Sea Shrimp**: Often referred to as the "cockroaches of the sea," these shrimp are hardy creatures that thrive in the abyss's extreme environment.

- **Black Swallower Fish:** This fish can swallow prey **larger than its body** thanks to its expandable stomach.

- **Hadal Snailfish:** Found at depths near the ocean's deepest points, these fish are among the deepest-living vertebrates discovered.

Fun Fact: Creatures in the abyss have evolved to survive without eyes, or with super-sensitive ones, using vibration or heat to sense their surroundings. Some even rely on **chemosynthesis** instead of photosynthesis for energy.

5. The Trenches (Hadalpelagic Zone)

- **Depth: 6,000 meters to 11,000 meters** (19,685 feet to 36,089 feet).

- **Characteristics:** The deepest, most isolated regions of the ocean, the trenches are extreme environments where the water is freezing, and the pressure is immense. The **Mariana Trench** is the most famous, where the water pressure is so intense it would crush most submarine equipment. Yet, despite the harsh conditions, the trenches are home to a number of species that are uniquely adapted to this extreme environment.

- **Marine Life:**

  - **Amphipods:** These small, shrimp-like creatures are surprisingly abundant in the trenches, where they survive by scavenging the remains of other creatures.

  - **Hadal Snailfish:** These fish thrive at the very bottom of the deepest trenches, surviving under the extreme pressure.

- **Giant Tube Worms:** These worms live near hydrothermal vents, where they rely on chemicals released by the vents to survive instead of sunlight.

Fun Fact: The **Challenger Deep** in the Mariana Trench is the deepest point on Earth, and it's so deep that if you dropped Mount Everest into it, the summit would still be **7,000 feet** underwater!

Five Major Oceans

1. The Pacific Ocean

The Pacific Ocean is the world's largest and deepest ocean, spanning over **63 million square miles**—more area than all the continents combined. It's home to some of the most iconic marine wonders, from vibrant coral reefs to mysterious trenches.

- **Notable Features:**
    - The **Mariana Trench**, the deepest point on Earth, plunging over 36,000 feet.
    - The **Great Barrier Reef**, the largest coral reef system, teeming with life.
    - The **Pacific Ring of Fire**, a region with active underwater volcanoes and frequent earthquakes.
- **Marine Life:**
    - **Sea Turtles:** Pacific waters host six of the world's seven species of sea turtles.
    - **Dugongs:** Gentle, slow-moving relatives of manatees.

- o **Whales**: Blue whales and humpback whales use the Pacific as their migratory highway.

- **Fun Fact**: The Pacific Ocean contains **more than half of all the water on Earth**, and it's so vast that it touches the shores of five continents!

2. The Atlantic Ocean

The Atlantic Ocean is the second-largest ocean, connecting the Americas to Europe and Africa. It's a vital artery for global trade and exploration, historically serving as a bridge between continents.

- **Notable Features**:
    - o The **Bermuda Triangle**, a region steeped in mystery.
    - o The **Mid-Atlantic Ridge**, the longest underwater mountain range.
    - o The **Sargasso Sea**, a region unique for its floating mats of sargassum seaweed.

- **Marine Life**:
    - o **Humpback Whales**: Known for their beautiful songs, they migrate through the Atlantic.
    - o **Seals**: From harp seals in the north to monk seals in the south.
    - o **Sea Turtles**: Atlantic beaches are critical nesting grounds for leatherback turtles.

- **Fun Fact**: The Atlantic is the saltiest ocean, and its Gulf Stream current is one of the strongest, influencing weather patterns across the globe.

3. The Indian Ocean

Known as the warmest ocean, the Indian Ocean is a treasure trove of biodiversity. Its tropical waters host rich ecosystems and play a crucial role in the monsoon weather patterns that affect billions of people.

- **Notable Features:**
    - The **Maldives**, a paradise of coral islands surrounded by turquoise waters.
    - The **Mozambique Channel**, home to one of the largest populations of whale sharks.
    - The **Indian Ocean Gyre**, a system of rotating currents.
- **Marine Life:**
    - **Dugongs**: Found grazing in the seagrass beds of the Indian Ocean.
    - **Whale Sharks**: The gentle giants of the ocean, commonly spotted here.
    - **Reef Fish**: Vibrant species that call coral reefs their home.
- **Fun Fact**: The Indian Ocean is the only ocean named after a country—India!

4. The Southern Ocean

Encircling Antarctica, the Southern Ocean is a realm of icy waters and dramatic landscapes. It plays a crucial role in regulating the Earth's climate by circulating cold water globally.

- **Notable Features:**

- - **Ice Shelves**: Massive floating platforms of ice, such as the Ross Ice Shelf.
  - **The Antarctic Circumpolar Current**, the strongest ocean current, connecting all major oceans.
  - **Krill Swarms**: Billions of tiny shrimp-like creatures form the foundation of the Antarctic food web.
- **Marine Life**:
  - **Emperor Penguins**: Iconic Antarctic residents.
  - **Leopard Seals**: Apex predators of the icy waters.
  - **Orcas**: These whales hunt in packs, working together to find prey.
- **Fun Fact**: The Southern Ocean's icy waters absorb more carbon dioxide than any other ocean, helping slow climate change.

5. The Arctic Ocean

The smallest and coldest ocean, the Arctic is a fragile ecosystem of ice, unique marine life, and extreme conditions.

- **Notable Features**:
  - **Floating Sea Ice**: A defining characteristic, shrinking rapidly due to global warming.
  - **Polynyas**: Areas of open water surrounded by ice, critical for marine mammals.
  - **The Beaufort Gyre**, an ocean current that traps freshwater under the ice.

- **Marine Life:**

  - **Narwhals:** Known as the "unicorns of the sea," these whales have long spiral tusks.

  - **Polar Bears:** Apex predators dependent on sea ice to hunt seals.

  - **Bowhead Whales:** Able to break through Arctic ice with their massive skulls.

- **Fun Fact:** The Arctic Ocean is warming faster than any other region on Earth, making it a critical focus for climate scientists.

Currents, Tides, and Waves: The Ocean in Motion

The ocean is never still—it's constantly moving, shaping our coasts, weather, and ecosystems through currents, tides, and waves.

Ocean Currents

Currents are like rivers within the ocean, carrying water, nutrients, and heat across the globe.

- **Surface Currents:**

  - Powered by wind and the Earth's rotation, these currents circulate warm and cold water around the planet.

  - **Example:** The **Gulf Stream**, which warms the coasts of Europe.

- **Deep Ocean Currents:**

  - Driven by differences in temperature and salinity, these currents create a "global conveyor belt" of water circulation.

- This system helps regulate Earth's climate by redistributing heat.

Fun Fact: Currents are so powerful that they can transport marine debris thousands of miles, creating "garbage patches" like the Great Pacific Garbage Patch.

Tides

Tides are the regular rise and fall of sea levels caused by the gravitational pull of the Moon and Sun.

- **High Tide:** When the water reaches its highest point.
- **Low Tide:** When the water recedes to its lowest level.
- **Spring Tides:** Extra-high tides that occur during a full or new moon, when the Sun, Moon, and Earth align.
- **Neap Tides:** Lower-than-normal tides that occur when the Sun and Moon are at right angles to Earth.

Fun Fact: In the Bay of Fundy in Canada, tides can rise over **50 feet**, the highest in the world!

Waves

Waves are formed by wind blowing over the surface of the water, but their size and power depend on factors like wind speed and duration.

- **Rogue Waves:** These massive, unexpected waves can tower over **60 feet** and are a danger even to large ships.
- **Tsunamis:** Giant waves caused by underwater earthquakes or volcanic eruptions, capable of traveling at speeds of up to **500 miles per hour**.

Ethereal Ray

Fun Fact: The largest wave ever recorded was **1,720 feet tall**, caused by a landslide in Lituya Bay, Alaska, in 1958!

Activity: Build Your Own Ocean Zones

Materials:

- Clear container or aquarium,
- Blue food coloring,
- Small figurines or toys (to represent marine life such as fish, turtles, squids, etc.),
- Markers or labels,
- Ruler (optional),
- Small stones or sand (optional, for the ocean floor).

Instructions:

1. **Layer the Zones:**
    - Start by filling your container with water.
    - Add a few drops of blue food coloring to represent the ocean water.
    - Divide the container into different layers using tape or markers, representing each ocean zone (Sunlight Zone, Twilight Zone, Midnight Zone, Abyss, and Trenches). You can either layer different shades of blue or place different items in each zone to indicate the depth.
    - Label each zone with a marker or paper label.
2. **Add Marine Life:**

- Place small figurines or toys representing marine creatures in each zone. For example:
    - **Sunlight Zone**: Colorful fish, dolphins, coral reefs, sea turtles.
    - **Twilight Zone**: Squid, lanternfish, anglerfish.
    - **Midnight Zone**: Giant squid, anglerfish, bioluminescent creatures.
    - **Abyss**: Deep-sea shrimp, sea cucumbers.
    - **Trenches**: Snailfish, amphipods, giant tube worms.

3. **Discuss the Zones:**
    - As you build the ocean, discuss the different conditions in each zone, like the amount of sunlight, temperature, and pressure. Talk about how the creatures have adapted to these extreme environments.
    - You can also discuss how the ocean zones are connected through currents, and how life thrives in such varied conditions.

4. **Ask Questions:**
    - What creatures would you expect to find in each zone? Why do you think they live there?
    - How do animals in the Twilight Zone find food without sunlight?
    - How do creatures in the Trenches survive the immense pressure?

Chapter 3:
Marine Life
- Creatures Big and Small

The ocean is home to an incredible array of creatures, some so tiny they're invisible to the naked eye, and others so large that they dwarf the biggest land animals.

These creatures live in a variety of environments, from the sun-drenched surface waters to the pitch-black depths of the ocean. Every marine animal has unique adaptations that allow it to survive in its specific environment, whether it's a microscopic plankton or the mighty blue whale.

Let's dive in and discover the creatures that call the ocean home!

Tiny but Mighty: The World of Plankton

While plankton may be small, they are incredibly important to the ocean's ecosystem.

These tiny creatures form the base of the food chain, feeding everything from small fish to the largest whales. The survival of most marine life depends on plankton's ability to thrive in the ocean.

- **Phytoplankton:**
  - These **microscopic plants** are the ocean's primary producers, meaning they use sunlight to create energy through photosynthesis. Just like trees on land, phytoplankton generate oxygen, and their tiny bodies are responsible for producing **over 50% of the oxygen** we breathe.
  - **Diatoms** and **coccolithophores** are common types of phytoplankton, each having a unique structure. Some diatoms have beautiful glass-like shells, while coccolithophores have chalky plates.

- Zooplankton:
  - These tiny animals are the consumers of the plankton world, feeding on phytoplankton. They range from tiny **copepods** (the most abundant marine creatures) to larger forms like **krill**, which are the primary food source for many larger animals.

- Some zooplankton, like **jellyfish larvae** and **salps**, are also incredibly important in marine ecosystems.

Fun Fact: Plankton are so numerous that if all the plankton in the ocean were gathered together, they would outweigh all the fish in the ocean combined!

The Fish of the Sea: From Tiny to Mighty

Fish are the ocean's most abundant vertebrates, and they come in all shapes and sizes. From the smallest species to the largest predators, each fish plays a key role in the ocean's food web.

- **Small Fish:**
  - **Clownfish:** These bright orange fish are famous for their symbiotic relationship with sea anemones. The clownfish are immune to the venom of the anemone's tentacles, and in exchange, they protect the anemone from predators.
  - **Sardines:** These small schooling fish are an essential food source for many marine animals, including seabirds, sharks, and whales. Sardines live in large schools that can contain thousands of fish, moving in synchrony to confuse predators.
  - **Parrotfish:** These beautifully colored fish are named for their beak-like teeth, which they use to scrape algae from coral reefs. Parrotfish play a

vital role in keeping coral reefs healthy by preventing algae from overgrowing.

- **Seahorses**: These quirky fish swim upright and are known for their unique appearance. Male seahorses carry the eggs in a pouch until they hatch, making them one of the few species where the male plays the role of "mother."

- **Big Fish:**
  - **Great White Shark**: One of the most feared predators in the ocean, the great white shark can grow up to **20 feet** long and weigh up to **5,000 pounds**. They are apex predators, keeping the populations of marine animals in balance.

  - **Manta Rays**: These graceful creatures have wingspans that can reach **up to 29 feet**. They glide through the water, feeding on plankton and small fish. Despite their massive size, manta rays are gentle giants that pose no harm to humans.

  - **Whale Sharks**: The **largest fish in the world**, whale sharks can grow over **40 feet** long. Despite their size, they are filter feeders, feeding on plankton and small fish. Whale sharks have a calm and docile nature, often swimming close to divers.

Fun Fact: The **whale shark** is often called the "gentle giant" because of its size and peaceful nature, despite being the largest fish on Earth.

Incredible Invertebrates: Creatures Without Backbones

Invertebrates make up about **90% of all marine species**, and many of these creatures are among the most fascinating. Without backbones, these animals rely on other features, like shells, exoskeletons, and unique adaptations to survive in the ocean's harsh environments.

- **Jellyfish:**

    o Jellyfish are ancient creatures, existing for over **500 million years**. Their bodies are 95% water, and they move by pulsating their bell-shaped bodies. Some species, The "immortal jellyfish" (Turritopsis dohrnii) can **revert to its juvenile form repeatedly**, making it biologically immortal.

    o Jellyfish use their long, stinging tentacles to capture prey and defend against predators. Some species, like the **box jellyfish**, have venom powerful enough to cause serious harm to humans.

    o **Fun Fact:** Jellyfish are so abundant in the ocean that they outnumber all the fish in some parts of the sea!

- **Octopuses:**

    - Octopuses are highly intelligent creatures known for their problem-solving skills. They are capable of using tools, camouflaging themselves, and even escaping from tight spaces.

    - With **eight arms** and **three hearts**, octopuses are incredible hunters, using their soft bodies to squeeze into small crevices to catch prey. They can also squirt ink to escape predators.

    - **Fun Fact**: Some octopuses have been known to open jars to access food inside—showing their impressive intelligence.

- **Crustaceans:**

    - **Crabs, lobsters, and shrimp** are all crustaceans, characterized by their hard exoskeletons. Crustaceans have to molt, or shed their shells, in order to grow.

    - **Mantis Shrimp:** Known for their incredible speed and strength, mantis shrimp can punch with the force of a bullet, stunning their prey in an instant. These shrimp can even break through aquarium glass with their powerful blows!

Fun Fact: The **coconut crab**, the world's largest terrestrial arthropod, can climb trees and crack open coconuts to eat!

Ethereal Ray
The Giants of the Sea: Mammals of the Ocean

While fish and invertebrates dominate the ocean's food chain, the ocean is also home to some of the largest creatures on Earth—marine mammals. These mammals are warm-blooded and breathe air, just like us, but they have evolved to live in the ocean.

- **Whales:**

  - **Blue Whale:** The blue whale is the largest animal to have ever lived—it's as long as two school buses and weighs as much as 30 elephants!

  - **Humpback Whales:** Known for their long migrations and complex songs, humpback whales travel thousands of miles each year, from cold feeding grounds to warm breeding grounds.

- **Sperm Whale**: These whales have the largest brains of any animal on Earth and can dive as deep as **10,000 feet** to hunt for giant squid.

- **Fun Fact**: The **blue whale's heart** is so big that a human could crawl through its arteries!

• **Dolphins:**

  - Dolphins are incredibly intelligent and social animals that use a wide range of vocalizations and body language to communicate.

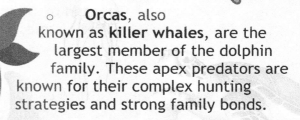

  - **Orcas**, also known as **killer whales**, are the largest member of the dolphin family. These apex predators are known for their complex hunting strategies and strong family bonds.

  - **Fun Fact**: Dolphins have been known to save humans from danger, guiding them to safety when lost at sea.

Creatures of the Deep: The Midnight and Abyssal Zones

The ocean's deeper zones—like the Midnight Zone, Abyss, and Trenches—are home to strange and fascinating creatures that have adapted to survive in extreme darkness, freezing temperatures, and immense pressure.

- **Anglerfish:**
    - The anglerfish uses a glowing lure to attract prey in the darkness of the Midnight Zone. Some species have jaws that can expand to swallow prey larger than their bodies.

- **Giant Squid:**
    - These elusive creatures can grow up to **40 feet** long and are among the most mysterious animals in the ocean. They live in deep water, where they prey on fish and other deep-sea creatures.

    Fun Fact: Giant squids can grow longer than a city bus! These deep-sea giants remain one of the ocean's greatest mysteries.

Chapter 4:
Coral Reefs
- The Rainforests of the Sea

Coral reefs are some of the most important ecosystems on Earth, often referred to as the **"rainforests of the sea."** Though they cover less than **0.1% of the ocean's surface**, they are home to about **25% of all marine species.**

Coral reefs provide food, shelter, and a breeding ground for countless marine organisms, making them a vital part of ocean health. However, these amazing ecosystems are increasingly under threat due to climate change, pollution, and human activity.

Let's explore what makes coral reefs so special and why they're so important for the health of our planet.

What Are Coral Reefs?

Coral reefs are like underwater cities built by tiny coral polyps. These animals stack their hard skeletons over time, creating colorful reefs full of life! As new polyps settle on top of the skeletons of older ones, the reef grows. Reefs can live for thousands of years, growing larger as time goes on.

Coral reefs are usually found in **warm, shallow waters** of the ocean, especially in tropical and subtropical regions. They grow best in **clear water** that is **not too cold**, typically found between **23.5° N and 23.5° S** of the equator.

Fun Fact: Some coral reefs are over 5,000 years old—older than the pyramids of Egypt!

The Importance of Coral Reefs

Coral reefs are incredibly valuable, not only to the marine life they support but also to humans. Let's break down the key roles they play in maintaining ocean health.

1. **Biodiversity Hotspot:**
   Coral reefs are one of the most **biodiverse ecosystems** on the planet. They provide food, shelter, and breeding grounds for a variety of species, including fish, mollusks, crustaceans, and marine mammals. It is estimated that **1 in 4 marine species** depend on coral reefs for survival, including **parrotfish**, **clownfish**, **sea turtles**, and **sharks**.

2. **Coastal Protection:**
   Coral reefs act as **natural barriers** that protect coastlines from the effects of **storm surges, high waves,** and **erosion**. They absorb the energy of waves and reduce the impact of storms, helping to protect coastal communities from hurricanes and tsunamis.

3. **Economic Value**:
   Coral reefs are a source of income for millions of people worldwide through **tourism, fishing**, and **medicine**. In some regions, coral reefs attract tourists who come to snorkel and dive in their clear, vibrant waters. Additionally, fish and seafood harvested from coral reefs provide food and livelihood for millions.

4. **Climate Regulation**:
   Coral reefs also help to regulate the ocean's temperature and are a **carbon sink**, meaning they absorb **carbon dioxide** from the atmosphere, helping to mitigate the effects of climate change.

Coral Reef Formation and Growth

Coral reefs grow in warm, shallow waters, where the water temperature is typically between **68°F and 82°F** (20°C to 28°C). To build their reefs, coral polyps need access to **sunlight** and **clean, clear water**. Coral and tiny algae called zooxanthellae are best friends! The algae give coral food, and the coral gives them a safe home.

As the polyps secrete their calcium carbonate skeletons, they build up the reef's structure. New polyps settle on top of older ones, creating the characteristic, often vibrant, reef formations.

Types of Coral Reefs

Reef Type	Description	Example
Fringing Reef	The most common type, grows directly along the shoreline. No large lagoon between reef and land.	Reefs in the Red Sea

Reef Type	Description	Example
Barrier Reef	Found farther from shore with a deep **lagoon** in between. Larger than fringing reefs.	**Great Barrier Reef** (Australia)
Atoll	A **ring-shaped** reef that forms around a **sunken volcanic island**, leaving a lagoon in the center.	**Maldives** Atolls
Patch Reef	Small, isolated reefs that grow in **shallow waters** between larger reefs.	Found in the **Caribbean**

Fun Fact: The Great Barrier Reef is the *largest coral reef system* on Earth, stretching over 1,400 miles!

The Creatures of the Reef

Coral reefs are teeming with life, and each zone within the reef supports different species. Let's explore some of the most interesting and important creatures that live on coral reefs:

1. **Coral Polyps:**

    o Coral polyps are the tiny animals that build the reef itself. Each polyp secretes a calcium carbonate skeleton, which over time, builds the massive reef structure.

    o Fun Fact: Some species of coral can grow up to **3 inches per year**, while others can grow up to **10 feet** in a single year, depending on the species and environmental conditions.

2. **Fish:**

    o Coral reefs are home to a vast variety of fish species, including **parrotfish**, **clownfish**,

lionfish, **snapper**, and **grouper**. Fish like **clownfish** live in a symbiotic relationship with **sea anemones**, where the anemone protects the fish with its stinging tentacles, and the clownfish offers food and protection.

- Fun Fact: **Parrotfish** are important because they help keep coral reefs clean by eating algae that might otherwise smother the corals.

3. **Sea Turtles:**

    - Several species of sea turtles, including the **green sea turtle** and **hawksbill sea turtle**, use coral reefs as feeding and nesting grounds. These turtles feed on algae, seagrasses, and small invertebrates found on the reef.

    - Fun Fact: **Hawksbill turtles** are important in controlling the growth of sponges that might otherwise overtake the corals.

4. **Crustaceans:**

    - **Crabs, lobsters, and shrimp** are abundant on coral reefs. Many species are scavengers, cleaning up debris and dead matter. Some, like the **mantis shrimp**, are fierce predators capable of delivering blows that can break glass aquarium tanks!

    - Fun Fact: **Cleaner shrimp** have a symbiotic relationship with fish, where they eat parasites and dead skin from the fish, providing a cleaning service.

5. **Mollusks:**

- Coral reefs are home to various mollusks, including **clams**, **oysters**, and **snails**. The **giant clam** can grow up to **4 feet** in length and weigh over **500 pounds**! These mollusks are important in reef ecosystems, helping filter the water and provide habitat for other species.

6. **Sharks and Rays:**

    - Coral reefs are also home to some of the ocean's most majestic predators, including **reef sharks** and **manta rays**. These animals help control the populations of smaller fish and invertebrates, maintaining the balance of the reef's ecosystem.

    - **Fun Fact: Manta rays** are gentle giants that glide through the water, feeding on plankton and small fish, and can have wingspans of up to **29 feet!**

Threats to Coral Reefs

Coral reefs are incredibly fragile and face many threats that are jeopardizing their survival. Climate change, pollution, overfishing, and destructive fishing practices are among the biggest dangers to coral reefs.

1. **Coral Bleaching:**

    - Coral bleaching occurs when the water temperature gets too high, causing the zooxanthellae algae to leave the coral. Without the algae, the coral loses its color and becomes stressed, making it vulnerable to disease and death.

    - **Fun Fact:** When coral is stressed by changes in temperature, light, or pollutants, it expels the

zooxanthellae algae, causing it to turn white—
this is called **coral bleaching**.

- When the ocean absorbs too much carbon dioxide, it becomes more acidic. This makes it harder for corals to build their skeletons—like trying to build a castle with melting sand!"

2. **Pollution:**

    - Pollution from plastic, oil, and agricultural runoff can smother coral reefs and poison marine life. The excess nutrients from fertilizers can also cause algae blooms, which block sunlight from reaching the coral.

    - **Fun Fact**: An estimated **8 million tons of plastic** enter the ocean every year, much of which ends up in coral reef ecosystems.

3. **Overfishing:**

    - Overfishing can deplete fish populations, disturb the balance of the ecosystem, and destroy coral reefs. Some destructive fishing practices, such as **blast fishing**, involve using explosives to stun fish, damaging the reef in the process.

    - **Fun Fact**: Some coral reefs have been overfished to the point where the marine life they used to support is now gone, leaving behind barren reefs.

Protecting Coral Reefs

Conservation efforts are crucial to ensuring the survival of coral reefs. Here's how we can help:

- **Marine Protected Areas (MPAs):** Establishing MPAs where fishing and human activity are limited can give coral reefs the space to recover and thrive.

- **Sustainable Fishing:** By supporting sustainable fishing practices, we can help prevent overfishing and protect the biodiversity of coral reefs.

- **Reducing Carbon Emissions:** Taking action on climate change can help prevent coral bleaching caused by rising ocean temperatures.

- **Pollution Reduction:** Reducing plastic waste and runoff from agriculture can help keep the oceans clean and healthy for coral reefs.

Activity: Create Your Own Coral Reef Ecosystem

Materials:

- Large piece of paper or cardboard,
- Colored markers, crayons, or colored pencils,
- Cut-out pictures of marine animals (from magazines or printed online),
- Glue.

Instructions:

1. Draw or trace a simple ocean scene on the paper or cardboard, with the coral reef as the focal point.
2. Add coral structures using the markers or crayons, making them colorful and diverse.
3. Cut out pictures of marine animals and place them in your reef—include fish, sea turtles, sharks, and invertebrates.
4. Discuss how these animals interact and depend on each other to maintain a healthy reef ecosystem.
5. Optional: Add labels to the creatures and explain their role in the ecosystem.

Chapter 5:
Ocean Exploration - Boldly Diving Deep

The ocean is one of the most mysterious places on Earth. Despite covering over **70% of the planet's surface**, we've only explored a tiny fraction of it—less than **20%**.

While space exploration often grabs the headlines, the exploration of our oceans is equally exciting and filled with wonder.

There are still so many mysteries beneath the waves, from uncharted underwater caves to unknown species living in the deep sea. But how do we explore these dark and distant places? Thanks to human ingenuity, we've created tools that allow us to venture deeper into the ocean than ever before.

In this chapter, we'll explore the evolution of ocean exploration, the early pioneers, and some of the exciting discoveries that continue to shape our understanding of the ocean. We'll also touch on how modern technology has expanded our reach, without diving too deeply into the specific tools, which we'll explore more in a later chapter.

The History of Ocean Exploration

Ocean exploration has evolved over the centuries, and significant milestones have been achieved in understanding the vast ocean world.

1. **The First Voyages**:

   - Early civilizations, such as the **Polynesians**, were pioneers in navigating the vast ocean using basic seafaring technology. They relied on stars, currents, and waves for navigation.

   - The **Phoenicians** and **Vikings** also made early explorations, traveling long distances by ship and mapping the Mediterranean and North Atlantic regions.

2. **The Age of Exploration**:

   - The **15th and 16th centuries** saw European explorers, including **Christopher Columbus** and **Ferdinand Magellan**, embarking on epic sea voyages to discover new lands and open up the world's oceans for future exploration.

   - In the **18th century**, **James Cook** made extensive maps of the Pacific Ocean, and his explorations led to the discovery of new lands, which helped open up scientific study of the ocean.

3. **The Birth of Oceanography**:

   - In the **19th century**, scientific expeditions began to take a more structured approach to ocean exploration. The **Challenger Expedition (1872-1876)** was one of the first major oceanographic missions, mapping the deep-sea trenches and documenting marine life. This marked the beginning of modern oceanography and a shift from simple exploration to scientific study.

Key Moments in Ocean Exploration

- ~3000 BCE – Polynesians navigate the Pacific using the stars.
- 15th-16th century – Columbus & Magellan explore and map coastlines.
- 1872-1876 – Challenger Expedition launches modern oceanography.
- 1960 – The *Trieste* reaches Challenger Deep.
- 2012 – James Cameron completes the first solo dive to the trench.

The Evolution of Ocean Exploration Technologies

While this chapter focuses on the milestones in ocean exploration, it is important to recognize how technology has dramatically expanded our ability to explore deeper and farther than ever before. Technologies like **sonar** and **submersibles** have played an integral role in the evolution of ocean exploration.

However, we'll go into more detail about these tools and **innovations** in Chapter 6, where we explore the **inventions** that have made modern ocean exploration possible. For now, let's look at how the early development of technology paved the way for what we can do today.

The Role of Early Ocean Exploration Technologies

1. **Sonar and Mapping:**

- Early explorers used rudimentary tools like **lead lines** to measure ocean depths, and in the 20th century, technologies like **sonar** allowed scientists to map the seafloor more accurately. Sonar uses sound waves to detect objects underwater and map underwater features. It revolutionized the way we understand the ocean floor and is still used today for exploring ocean depths.

2. **The First Submersibles and ROVs:**

    - The development of the first **submersibles** in the mid-20th century allowed humans to venture into the deep ocean, reaching depths never before possible. These vehicles helped scientists discover new species, underwater landscapes, and shipwrecks.

    - **Remotely Operated Vehicles (ROVs)** were introduced in the 1970s to perform deep-sea exploration tasks without human involvement. These unmanned vehicles allowed scientists to access even the most dangerous parts of the ocean.

Modern-Day Ocean Exploration and Exciting Discoveries

Ocean exploration continues to reveal new and exciting discoveries that challenge our understanding of marine life, the ocean floor, and the planet's history.

1. **The Mariana Trench:**

    - The **Mariana Trench** is the deepest part of the ocean. In 1960, the *Trieste* became the first human-occupied vehicle to reach the Challenger Deep in the Mariana Trench. More recently, in

2012, filmmaker James Cameron made the first solo dive to the bottom!

- Today, scientists use deep-sea robots like *Alvin* and remotely operated vehicles (*ROVs*) to explore ocean depths that humans can't reach. These high-tech machines help us uncover shipwrecks, study hydrothermal vents, and discover new species!

2. **Hydrothermal Vents:**

   - The discovery of **hydrothermal vents** in the 1970s opened a new chapter in our understanding of life on Earth. These underwater geysers, where hot water rich in minerals is expelled from the seafloor, host entire ecosystems that survive without sunlight, relying on **chemosynthesis**.

3. **Shipwrecks and Sunken Cities:**

   - Submersibles and ROVs have uncovered ancient shipwrecks, including the **Titanic**, and submerged cities like **Dwarka** off the coast of India. These discoveries have provided valuable insights into history and the maritime cultures of the past.

Looking to the Future: The Next Frontier in Ocean Exploration

While we've made significant advances in ocean exploration, there is still so much left to discover. The ocean holds **vast, unexplored areas** that remain a mystery. With new technologies, like **autonomous underwater vehicles (AUVs)** and **satellite mapping**, we are moving closer to uncovering the ocean's greatest secrets.

Fun Fact: Scientists believe that **90%** of the ocean remains unexplored, and that the deep ocean may harbor millions of undiscovered species.

Activity: Design Your Own Ocean Exploration Mission

Materials:

- Paper, colored pencils, and markers,
- Reference pictures of submersibles, ROVs, and underwater creatures.

Instructions:

1. Imagine you are a marine explorer preparing for an exciting mission to the deep sea.

2. Draw and design your own **submersible** or **ROV**, making sure it's equipped with the right tools for the job (camera, sample collectors, lights).

3. Think about where you would explore (the **Mariana Trench**, a coral reef, or an underwater volcano) and the creatures you might encounter. Draw the creatures and the environment you expect to find.

4. Present your mission to others, explaining your exploration plan and what discoveries you hope to make.

Conclusion: A Never-Ending Journey of Discovery

Ocean exploration is an exciting and ongoing adventure. As technology advances, we are reaching new depths and discovering secrets that have been hidden for millions of years. There is still so much to learn about our oceans—who knows what amazing discoveries await?

Chapter 6:
Incredible Ocean Inventions

The ocean has always been a gateway to new lands, cultures, and discoveries.

Throughout history, technological innovations have allowed people to sail across vast distances, explore unknown continents, and eventually unlock the mysteries of the deep sea. From the first **seafaring vessels** to modern **submersibles** and **underwater drones**, ocean technology has shaped human history and exploration.

In this chapter, we will dive into both **historical** and **modern ocean technologies**—from ancient sailing innovations that allowed early explorers to cross the seas to cutting-edge technologies that help us explore and protect the oceans today. These inventions have not only advanced ocean exploration but also helped us better understand and conserve the marine world.

Early Technologies for Ocean Travel

Before we could explore the depths of the ocean, humans needed to learn how to **sail across it**. The invention of sailing vessels, navigational tools, and techniques opened the doors to exploration and trade, connecting distant lands and peoples.

The First Sailing Vessels

The earliest humans navigated across seas using simple **rafts** and **canoes** made from materials like wood, animal skins, and reeds. These early vessels were propelled by paddles and the current.

- **Rafts and Canoes:**
  Early **Polynesian navigators** were known for their exceptional skills in using **outrigger canoes** to travel long distances across the Pacific Ocean. They navigated using the stars, winds, and ocean currents, reaching distant lands like **Hawaii**, **New Zealand**, and **Easter Island**.

Fun Fact: The **Polynesians** sailed thousands of miles across the Pacific, guided only by their deep understanding of the ocean's currents and natural signs.

The Invention of Sails: The Birth of Long-Distance Travel

The invention of the **sail** revolutionized ocean travel, allowing ships to harness the wind and travel faster and more efficiently.

- **The Early Sailboat:**
  The **Egyptians** were among the first to use sails around **5,000 years ago**, navigating the Nile River and the Mediterranean Sea. Early sailboats used **square sails**,

which were effective for catching the wind but didn't allow for great maneuverability.

Fun Fact: Early sailboats used **reed boats**, and over time, sailors developed **triangular sails**, which allowed ships to sail against the wind and made longer voyages possible.

Navigational Tools: Mapping the Seas

As sailors ventured farther from home, they needed tools to navigate the vast and unknown oceans. This led to the development of key **navigational tools**.

- **The Compass:**
  The invention of the **magnetic compass** in **ancient China** was revolutionary. It allowed sailors to determine direction even on cloudy days or at night, improving navigation.

Fun Fact: The compass was such a game-changer that it's still used in modern navigation today!

- **The Sextant:**
  In the 18th century, the **sextant** was invented. This tool helped sailors measure the angle between the horizon and celestial bodies, allowing them to calculate their **latitude** and **longitude** accurately.

The Age of Exploration: Ships that Changed the World

During the **Age of Exploration** (15th to 17th centuries), the invention of powerful ships allowed explorers to travel across the world, paving the way for **global trade** and **discovery**.

- **The Caravel:**
  The **caravel**, developed by the **Portuguese** in the 15th century, was a light, fast ship with **triangular sails**, making it ideal for exploring distant lands. Explorers

like **Christopher Columbus** used caravels on their historic voyages.

Fun Fact: The **Santa Maria**, Columbus's flagship, was a caravel!

- **The Galleon:**
  The **galleon**, a large multi-decked ship, was used for both trade and warfare in the 16th and 17th centuries. Galleons were vital in the **Spanish treasure fleets**, bringing gold and silver from the Americas to Europe.

Fun Fact: Galleons could carry massive cargoes, including **silk, spices, and precious metals** from across the globe!

2. The Development of Modern Ocean Exploration Technologies

With the invention of tools like the **aqualung** and **submersibles**, humans began to explore not only the surface of the oceans but also the deep, uncharted regions of the seas.

The Birth of Scuba Diving

In the 20th century, the invention of the **aqualung** allowed divers to explore the oceans for extended periods without being tethered to the surface.

- **The Aqualung:**
  In **1943**, **Jacques Cousteau** and **Émile Gagnan** invented the **aqualung**, the first self-contained underwater breathing apparatus (SCUBA). This invention revolutionized ocean exploration, allowing divers to explore freely and stay submerged for hours.

Fun Fact: Jacques Cousteau's documentaries popularized SCUBA diving and brought the beauty of the underwater world into the homes of millions of people worldwide!

Submersibles: Unlocking the Depths

To reach the ocean's deepest, most remote regions, submersibles were developed—specialized vessels capable of withstanding immense underwater pressure.

- **The Bathyscaphe:**
  In **1960**, the **Trieste**, a bathyscaphe designed by **Auguste Piccard**, became the first vehicle to reach the bottom of the **Mariana Trench**—the deepest part of the ocean. The Trieste reached a depth of over **35,000 feet**.

Fun Fact: The **Mariana Trench** is so deep that **Mount Everest** could fit inside it, with the peak still over **7,000 feet underwater!**

- **Modern Submersibles:**
  Today, submersibles can explore the ocean's deepest regions, reaching depths of over **36,000 feet**. They are equipped with cameras, robotic arms, and lights to collect samples and capture images of the ocean floor.

Remotely Operated Vehicles (ROVs)

To explore places too dangerous for humans, **ROVs** were developed. These unmanned vehicles are controlled from the surface and can be used for deep-sea exploration, underwater archaeology, and environmental research.

- **How ROVs Work:**
  ROVs are equipped with cameras, lights, and robotic arms. They are especially useful for tasks like collecting samples from the seafloor, mapping underwater ecosystems, and exploring shipwrecks.

Autonomous Underwater Vehicles (AUVs)

- **Scientists are developing advanced Autonomous Underwater Vehicles (AUVs)**—robots that explore the ocean without needing a human pilot! These high-tech machines help us map the seafloor, monitor marine life, and even clean up pollution.

- **How AUVs Work:**
  AUVs are designed to gather data, measure ocean currents, map the seafloor, and monitor marine ecosystems. These vehicles are revolutionizing how we study the deep sea.

Ocean Cleanup Technologies

With the rise of ocean pollution, new technologies have been developed to **clean up** plastic waste and protect marine ecosystems. Projects like **The Ocean Cleanup Project** use innovative methods to remove debris from the ocean.

- **The Ocean Cleanup Project:**
  This ambitious project uses floating barriers and filtration systems to capture and remove plastic debris from the **Great Pacific Garbage Patch**.

3. Future Technologies in Ocean Exploration

The future of ocean exploration is filled with possibilities. As technology continues to evolve, new tools and innovations will help us explore, protect, and understand the ocean more deeply.

- **Underwater Drones:**
  Underwater drones are becoming increasingly important for marine research and conservation. These autonomous devices can collect real-time data, take

high-resolution photos, and explore remote ocean locations.

- **Robotic Submarines:**
  Future **robotic submarines** will allow scientists to explore the deepest, most inaccessible parts of the ocean. These highly advanced vehicles will have the ability to survive extreme conditions and provide data on ocean health, marine species, and the ocean floor.

Old Technology	Modern Version	Purpose
Rafts & Canoes	Motorized Ships	Ocean Travel
Compass & Sextant	GPS & Satellite Navigation	Finding Direction
Bathyscaphe *Trieste* (1960)	Triton 36000/2 (2019)	Deep-Sea Exploration
Early Sonar (1910s)	Multibeam Sonar (Today)	Ocean Mapping
Basic Diving Gear	SCUBA & Exosuits	Human Exploration

Activity: Design Your Own Ocean Exploration Invention

Materials:

- Paper, colored markers, pencils, scissors
- Reference pictures of early ships, modern submersibles, and underwater robots

Instructions:

1. Imagine you are an inventor tasked with creating a new tool for exploring the ocean.

2. Design your invention—whether it's a submersible, underwater drone, or a new piece of technology.

3. Draw your invention and label its key features.

4. Present your invention to others, explaining how it would help scientists explore the ocean more effectively.

Conclusion: The Ocean Awaits

From ancient sailing vessels to cutting-edge technologies, ocean exploration has come a long way. Thanks to these incredible inventions, we've been able to uncover the mysteries of the deep and work toward protecting our oceans. The future of ocean exploration is bright, and as technology continues to evolve, there's still so much more to discover beneath the waves.

Chapter 7: Ocean Phenomena and Natural Wonders

The ocean is not just a vast body of water—it is a realm filled with **mysterious phenomena** and **incredible wonders** that continue to captivate scientists and explorers.

While much has been discovered, there are still countless **secrets** hidden beneath the waves. From glowing creatures and underwater volcanoes to strange waves and currents, the ocean is full of phenomena that defy our understanding and challenge the boundaries of science.

In this chapter, we will explore some of the most awe-inspiring and baffling phenomena of the ocean, delving into the mysteries that make the seas so fascinating and unpredictable.

Bioluminescence: The Glow of the Ocean

One of the most magical and mysterious phenomena in the ocean is **bioluminescence**—the ability of living organisms to produce light. This glow is not just beautiful but also helps marine creatures survive in the deep sea.

What Is Bioluminescence?

- Bioluminescence occurs when a chemical reaction inside an organism produces light.
- The enzyme **luciferase** reacts with a molecule called **luciferin**, creating a glow.
- Marine creatures use this for different reasons:
    - Some **attract prey**.
    - Others **scare off predators**.
    - Some **find mates** in the dark ocean depths.

Where Does Bioluminescence Occur?

- Bioluminescent creatures are most commonly found in the **Midnight Zone**, where sunlight does not reach.
- Some examples include:
    - **Firefly squid**
    - **Deep-sea jellyfish**
    - **Anglerfish**
- **Glowing plankton** near the surface can create sparkling waves at night.

Fun Fact: Some fireflies glow in the same way as bioluminescent sea creatures!

Rogue Waves: The Ocean's Killer Waves

A rogue wave is a massive, unexpected wave that can tower over surrounding waves and pose a serious danger to ships.

What Are Rogue Waves?

- Rogue waves are at least **twice the height of surrounding waves**.
- They can form when:
  - Smaller waves merge into a giant wave.
  - Powerful ocean currents interact and create sudden energy bursts.
- Some rogue waves reach **over 100 feet (30 meters) high**, capable of:
  - Capsizing ships
  - Damaging oil rigs

Why Are Rogue Waves Dangerous?

- Rogue waves appear **suddenly**, even in calm waters, making them unpredictable.
- Sailors and scientists still study them to understand their formation.

Fun Fact: Rogue waves are sometimes called "freak waves" because they seem to appear out of nowhere, defying normal ocean patterns.

Underwater Volcanoes and Hydrothermal Vents

Beneath the ocean's surface, underwater volcanoes and **hydrothermal vents** reshape the seafloor, releasing heat, gases, and minerals that create unique deep-sea ecosystems.

Underwater Volcanoes

- These volcanoes form along **tectonic plate boundaries**.
- When magma escapes from the Earth's crust, it cools and forms new land.
- Some of the largest volcanoes, such as **Mauna Kea in Hawaii**, are actually underwater.

Hydrothermal Vents

- Hydrothermal vents are **underwater geysers** found along mid-ocean ridges.
- They spew out **hot, mineral-rich water**, creating deep-sea ecosystems.
- Creatures that live here, such as:
    - **Tube worms**
    - **Giant clams**
    - **Shrimp**
- These organisms use **chemosynthesis**, a process that converts chemicals into energy instead of sunlight.

Fun Fact: Hydrothermal vents support life that doesn't rely on sunlight, challenging scientists' ideas of where life can exist.

Ocean Currents and the Great Conveyor Belt

The ocean is constantly moving due to powerful currents driven by wind, temperature differences, and the Earth's rotation. These currents regulate the **climate** and support **marine life**.

What Are Ocean Currents?

- Ocean currents are continuous movements of seawater.
- They help:
    - **Regulate temperatures**
    - **Distribute nutrients**
    - **Aid in the migration of marine animals**

The Great Conveyor Belt

- One of the most important ocean currents is the **thermohaline circulation**, also called the **Great Conveyor Belt**.

- It moves:
  - **Warm water** from the equator to the poles.
  - **Cold water** from the poles back to the equator.
- This circulation **helps regulate Earth's climate**.

Fun Fact: The Great Conveyor Belt takes about **1,000 years** to complete one full cycle around the globe!

Ocean Mysteries
The Bloop – A Deep-Sea Mystery

- In **1997**, scientists recorded a powerful underwater sound known as **The Bloop**.
- Some believed it came from a **massive sea creature**.
- Scientists now think it was caused by **the cracking of an Antarctic ice shelf**.

The Bermuda Triangle: A Mysterious Region

- The **Bermuda Triangle** is located between **Miami, Bermuda, and Puerto Rico**.
- Over the years, ships and aircraft have disappeared under strange circumstances.
- Theories range from:
  - **Magnetic anomalies**
  - **Extreme weather**
  - **Human navigation errors**
- Scientists believe most cases have logical explanations.

Fun Fact: The Bermuda Triangle has inspired countless books, movies, and documentaries, making it a modern legend.

Famous Shipwrecks: The Mystery of the Titanic

- The **RMS Titanic** was one of the most famous shipwrecks in history.

- It sank in **1912** after hitting an iceberg in the North Atlantic.
- The wreck was discovered in **1985** by a team using sonar and deep-sea submersibles.
- The Titanic's sinking led to improvements in **maritime safety laws**.

Fun Fact: The Titanic lies over **12,400 feet (3,800 meters) below the surface!**

Conclusion: The Ocean's Endless Mysteries

The ocean is full of wonder, from glowing creatures and rogue waves to deep-sea volcanoes and uncharted regions. As technology advances, we are uncovering more about these hidden worlds—but many secrets remain. The mysteries of the ocean remind us how much there is still to explore, study, and protect.

Activity: Create Your Own Ocean Phenomenon

Materials:

- Paper, colored markers, crayons, or pencils
- Reference pictures of ocean phenomena (bioluminescence, rogue waves, hydrothermal vents, etc.)

Instructions:

1. Imagine you've discovered a new ocean phenomenon that no one has ever seen before.
2. Design what it looks like and describe how it works.
3. Draw your ocean phenomenon and explain its significance in the ocean ecosystem.

Ethereal Ray

🐟 **Who knows what other ocean mysteries we will discover in the future?**

Chapter 8:
Ocean Careers
- Your Path to Marine Adventures

Have you ever dreamed of exploring coral reefs, studying sharks, or saving endangered sea turtles? The ocean offers a wide range of fascinating careers for those who are passionate about marine life, conservation, and exploration.

From being a marine biologist to designing underwater robots, there are countless ways to turn your love for the ocean into a lifelong adventure.

In this chapter, we'll explore some of the most exciting ocean-related careers, what they involve, and how you can prepare to join the ranks of ocean explorers, scientists, and conservationists.

1. Marine Biologist: Studying Life Beneath the Waves

A career as a marine biologist is one of the most popular paths for ocean lovers. Marine biologists study the plants, animals, and ecosystems of the sea, from tiny plankton to massive whales.

What Marine Biologists Do:

- Conduct research on marine species and ecosystems.
- Monitor the health of coral reefs, mangroves, and other vital habitats.
- Protect endangered species, like sea turtles, sharks, and manatees.
- Study how climate change, pollution, and overfishing affect marine life.

Where They Work:

- Universities, research institutions, aquariums, and conservation organizations.
- Many marine biologists work in the field, traveling to remote locations for underwater research.

Fun Fact: Marine biologists often use **SCUBA diving gear** to study coral reefs up close!

2. Oceanographer: Unlocking the Secrets of the Sea

Oceanographers study the ocean's physical, chemical, geological, and biological aspects, helping us understand how the ocean works and its role in Earth's climate.

Types of Oceanographers:

1. **Physical Oceanographers:** Study waves, tides, and currents.
2. **Chemical Oceanographers:** Analyze the chemical composition of seawater.

3. **Geological Oceanographers:** Study the ocean floor and underwater mountains.
4. **Biological Oceanographers:** Focus on marine life and ecosystems.

What Oceanographers Do:

- Explore the deep sea using **submersibles and ROVs**.
- Analyze water samples to understand ocean health.
- Study how ocean currents affect weather and climate.

Fun Fact: Oceanographers discovered the **Great Ocean Conveyor Belt**, a global system of currents that regulates Earth's climate!

3. Marine Engineer: Designing and Building Ocean Technology

Marine engineers design and build the tools, vehicles, and structures needed to explore and utilize the ocean.

What Marine Engineers Do:

- Design and construct **submarines, ROVs, and AUVs**.
- Develop technology for **offshore wind farms, tidal turbines, and desalination plants**.
- Improve **diving gear and underwater habitats** for researchers and explorers.

Fun Fact: Marine engineers helped design the **Deepsea Challenger**, the submersible that reached the bottom of the **Mariana Trench in 2012!**

4. Underwater Archaeologist: Exploring History Beneath the Sea

Underwater archaeologists combine a love of history with a passion for diving. They study shipwrecks, sunken cities, and other submerged cultural treasures.

What Underwater Archaeologists Do:

- Search for and excavate **shipwrecks**, such as the Titanic and ancient Roman vessels.
- Study artifacts found underwater, like pottery, tools, and weapons.
- Use sonar and underwater drones to **map and explore sunken sites**.

Fun Fact: Some underwater archaeologists have discovered entire **sunken cities**, like **Dwarka off the coast of India!**

5. Marine Conservationist: Protecting Our Oceans

Marine conservationists work to safeguard marine environments from threats like overfishing, pollution, and climate change.

What Marine Conservationists Do:

- Lead efforts to clean up plastic from the ocean.
- Advocate for policies that **protect marine life and habitats**.
- Educate communities about **sustainable fishing** and reducing pollution.

Fun Fact: Marine conservationists helped establish **Marine Protected Areas (MPAs)** to safeguard coral reefs and endangered species worldwide.

6. Aquarist: Caring for Ocean Creatures in Captivity

Aquarists take care of marine animals in aquariums and marine parks, ensuring they remain healthy and engaged.

What Aquarists Do:

- Feed, clean, and monitor the health of marine animals.
- Design and maintain **realistic habitats** for ocean creatures.
- Educate visitors about **marine life and conservation**.

Fun Fact: Aquarists work with sharks, dolphins, and sea otters, providing **enrichment activities** to keep them active and healthy!

7. Ocean Photographer/Filmmaker: Capturing the Beauty of the Sea

Ocean photographers and filmmakers use their skills to showcase marine life and ocean environments, raising awareness about conservation.

What Ocean Photographers Do:

- Dive into underwater environments to capture stunning images.
- Create documentaries on **coral reefs, marine species, and climate change**.
- Travel to remote locations to **document the effects of climate change on the ocean**.

Fun Fact: Jacques Cousteau was one of the first to film the underwater world, inspiring millions to care about the ocean!

8. Other Ocean-Related Careers

The ocean offers many unique career paths, including:

- **Marine Policy Advisor:** Works on laws that protect the ocean.
- **Diving Instructor:** Teaches others how to explore underwater safely.
- **Coastal Engineer:** Designs structures to protect coastlines from erosion and storms.
- **Marine Veterinarian:** Cares for injured or sick marine animals.

How to Start Your Ocean Career

If you're interested in working with the ocean, follow these steps:

1. **Learn about the ocean:** Read books, watch documentaries, and explore tide pools or aquariums.
2. **Study science:** Focus on **biology, physics, and chemistry**—many ocean careers require a strong science background.
3. **Get involved:** Volunteer for **beach cleanups, marine conservation projects, or aquarium programs.**
4. **Get SCUBA certified:** Many ocean jobs require **diving experience.**
5. **Stay curious:** The ocean is full of mysteries, and curiosity will help you succeed in any marine career!

Activity: Imagine Your Future Ocean Career

Materials:

- Paper, colored markers, crayons, or pencils

Instructions:

1. Think about which ocean career excites you the most.

2. Draw yourself in that career—whether you're a marine biologist studying whales, a conservationist cleaning up plastic, or an aquarist caring for dolphins.
3. Write a short paragraph about what you would do in your dream ocean career and why it's important to you.

Conclusion: Make a Difference Beneath the Waves

The ocean offers endless opportunities for **adventure, discovery, and making a positive impact.** Whether you dream of studying marine life, exploring shipwrecks, or designing underwater technology, there's a career waiting for you. By working to understand and protect the ocean, you can help ensure its wonders remain for future generations to enjoy.

Chapter 9: Legends of the Sea - Mythology and Folklore

The ocean has long been a source of fascination, inspiring myths, legends, and folklore across every culture touched by its waves.

These stories reflect both the beauty and terror of the seas, portraying it as a realm of magical beings, fearsome monsters, and lost civilizations. Whether based on real phenomena or pure imagination, these tales continue to captivate us.

In this chapter, we'll explore more **ocean legends**, from lost lands like Lemuria to mysterious vanishing islands and mythical creatures. We'll dive into folklore across continents and uncover the possible truths behind the tales.

Mermaids and Sirens: Enchantresses of the Sea

Mermaids and sirens have captured imaginations for centuries, appearing in myths as both alluring and dangerous. Let's explore their fascinating duality and how they've evolved over time.

- **Benevolent Mermaids:**
    - In **Caribbean folklore**, mermaids are seen as guardians of the sea. Fishermen often leave offerings to the mermaids, asking for safe journeys and abundant catches.
    - In **Native American tales**, mermaid-like beings called **Water Spirits** are protectors of lakes and rivers, guiding those who respect nature.
- **Dark and Dangerous Sirens:**
    - While Greek **Sirens** lured sailors to shipwrecks with their songs, **Russian Rusalki** are water spirits who seek revenge on humans who wronged them.

Did You Know? Some mermaid myths may be connected to the manatee's unique tail, which resembles a mermaid's fin when seen from a distance.

2. Lemuria: The Lost Land Beneath the Pacific

Lemuria, sometimes called **Mu**, is a mythical sunken continent believed to have existed in the **Pacific Ocean**.

- **The Legend of Lemuria:**
    - Spiritualists describe Lemuria as a paradise where its inhabitants lived in harmony with

nature. Lemurians were thought to possess great wisdom and advanced technologies.

- According to legend, Lemuria sank due to rising waters, with survivors said to have fled to places like Mount Shasta in California.

- **Possible Inspiration:**
    - Some believe Lemuria could have been inspired by the **subduction zones** in the Pacific, where tectonic plates collide and sink.

Fun Fact: Lemuria often appears in modern literature and fantasy, inspiring novels, video games, and films about ancient lost civilizations.

3. Atlantis: The Great Lost Civilization

Atlantis remains one of the most enduring tales of a sunken world. Its story comes from **Plato**, who described it as an advanced and wealthy society destroyed by divine punishment.

- **Possible Locations:**
    - **Santorini**, with its volcanic eruption, and **Doggerland**, a prehistoric land in the North Sea, are often cited as potential inspirations.
    - Others suggest Atlantis could be purely allegorical, symbolizing the dangers of hubris.

Fun Fact: Expeditions to find Atlantis continue today, with some enthusiasts even suggesting locations in the **Caribbean** or **Antarctica**!

4. Sea Serpents: Guardians and Terrors

Sea serpent myths are nearly universal, with creatures described as enormous, snake-like, and terrifying.

- **Modern Sightings**:
    - Reports of **Loch Ness Monster**-like creatures in oceans around the world continue to spark debate.
    - Historical sightings of **oarfish** may explain many sea serpent legends. These rarely seen creatures are among the ocean's longest fish and have a serpent-like appearance.

Did You Know? In Norse mythology, **Jörmungandr**, the Midgard Serpent, is so large that it circles the Earth, biting its own tail.

5. Ghost Ships: Haunted Vessels of the Sea

Tales of **ghost ships** abound, blending the supernatural with real-life maritime mysteries.

- **The Lady Lovibond**:
  Legend says this ship wrecked on the **Goodwin Sands** in 1748, doomed to reappear every 50 years. Sightings were reported in **1798**, **1848**, and **1948**, though no wreckage has been found.

- **Modern Ghost Ships**:
  Real-life ghost ships, like the **MV Lyubov Orlova**, a derelict cruise liner set adrift, spark modern intrigue. It was spotted drifting off the North Atlantic before vanishing entirely.

6. The Vanishing Island of Hy-Brasil

In Irish mythology, **Hy-Brasil** is a legendary island said to appear only once every seven years. Often shrouded in mist, it's believed to be a land of great wealth and knowledge.

- **Explorer Sightings**:
    - In 1497, explorer **John Cabot** claimed to have seen Hy-Brasil on his transatlantic journey.
    - Some believe it could have been a mirage or an actual volcanic island that has since disappeared.

Fun Fact: Hy-Brasil has been linked to UFO sightings, with some suggesting it might have extraterrestrial origins.

7. The Ningen: A Modern Legend

The **Ningen**, a modern cryptid from Japanese folklore, is described as a **human-like creature** living in the icy waters of the **Antarctic Ocean**.

- **Sightings and Descriptions**:
    - Described as **whale-sized**, with pale white skin and human-like facial features, the Ningen has been reported by sailors and researchers.
    - Some believe it could be an undiscovered species of marine mammal, while others suggest it's a hoax.

Did You Know? The name "Ningen" means **human** in Japanese, emphasizing its eerie resemblance to us.

8. Charybdis and Scylla: The Greek Sea Monsters

In ancient Greek mythology, sailors navigating the **Strait of Messina**, the narrow passage between Italy and Sicily, faced the perils of **Charybdis** and **Scylla**—two monstrous threats that made the journey treacherous. These myths symbolize the dangers of navigating real-life oceanic hazards.

- **Charybdis:**
  A monstrous whirlpool that sucked in everything nearby, Charybdis was said to be the daughter of Poseidon, cursed by Zeus to become a sea monster. Sailors who ventured too close risked being swallowed whole.

- **Scylla:**
  Across from Charybdis lived Scylla, a creature with **six serpent-like heads,** each capable of snatching a sailor from passing ships. She was once a beautiful nymph cursed by the sorceress **Circe** into her monstrous form.

- **The Real-Life Connection:**
  The **Strait of Messina** is known for its strong currents and whirlpools, which may have inspired these legends. Navigating the strait required skill and courage, much like the mythical heroes who faced Charybdis and Scylla.

Fun Fact: The phrase **"between Scylla and Charybdis"** is used to describe being caught between two equally dangerous situations!

9. The Umibōzu: Japan's Shadowy Sea Spirit

In Japanese folklore, the **Umibōzu** is a ghostly figure that haunts sailors and fishermen. Known for its ominous presence and unpredictable nature, the Umibōzu embodies the mysterious and sometimes terrifying power of the sea.

- **Appearance:**
  The Umibōzu is often described as a **giant, dark figure** rising from the water, with a smooth, round head

resembling a monk's. Some legends say it has **no visible features**, while others describe glowing eyes or an unsettlingly human-like face.

- **Behavior:**
  Legends say the Umibōzu appears suddenly, especially during calm nights, and demands an offering from sailors—often a barrel. If the sailors fail to appease it, the spirit creates massive waves or capsizes the ship.

- **Possible Inspirations:**

  o Rogue waves, sudden weather changes, or the eerie stillness of the sea on calm nights may have contributed to sightings of the Umibōzu.

  o Its monk-like appearance may relate to Japan's tradition of connecting spirits with Buddhist imagery.

Fun Fact: The Umibōzu is still referenced in modern Japanese pop culture, appearing in anime, movies, and even video games.

10. St. Elmo's Fire: Mysterious Ocean Lights

For centuries, sailors have reported seeing strange, glowing blue or green lights atop their ship's masts during storms. Known as **St. Elmo's Fire**, this phenomenon was often interpreted as a **divine sign** or a warning from the heavens.

- **What Is St. Elmo's Fire?**
  It's an electrical phenomenon caused by a discharge of **static electricity** during thunderstorms. The glowing effect appears as bluish flames or sparks and is named after **St. Erasmus (Elmo)**, the patron saint of sailors.

- **Historical Beliefs**:
  Sailors in ancient times believed that St. Elmo's Fire

was a sign of **protection** from their saint, reassuring them that their ship would survive the storm. However, some also saw it as an omen of impending doom.

- **Modern Understanding**:
  Scientists explain that the phenomenon occurs when the air becomes highly charged with electricity, creating a plasma-like glow.

Fun Fact: While it's most commonly seen on ship masts, St. Elmo's Fire can also occur on airplane wings and other pointed objects.

11. The Makara: India's Water Guardian

The **Makara** is a mythical sea creature in Hindu and Buddhist traditions, revered as a **symbol of protection** and **abundance**. It is often associated with deities like **Varuna**, the god of the sea, and **Ganga**, the river goddess.

- **Appearance**:
  The Makara is depicted as a hybrid creature, with the body of a fish and the head of a crocodile, elephant, or other animals. Its unique and varied features symbolize the interconnectedness of aquatic and terrestrial life.

- **Cultural Significance**:

  - In Hinduism, the Makara is the **vahana (vehicle)** of Varuna and Ganga. Statues of Makara often guard temple entrances and symbolize protection from evil.

  - In Buddhist art, the Makara appears as a decorative motif on temple roofs and gateways, representing the transition between the earthly and spiritual realms.

Fun Fact: In modern times, the Makara is celebrated in art and architecture, appearing on bridges, gates, and even jewelry designs.

12. The Leviathan: A Biblical Sea Monster

The **Leviathan**, mentioned in ancient Hebrew texts and the Bible, is a mighty sea serpent symbolizing chaos and destruction. It is often portrayed as a creature so massive that it could churn the seas with a single movement.

- **Biblical References:**
  The Leviathan is described in the **Book of Job** as an untamable beast, feared by all and revered as a symbol of divine power. In some interpretations, it represents the chaos that existed before the world was ordered by God.

- **Modern Symbolism:**
  The Leviathan has become a cultural symbol of overwhelming power and danger. It appears in literature, from **Thomas Hobbes's Leviathan** to fantasy works inspired by its legendary might.

Fun Fact: Some scholars believe the Leviathan was inspired by real animals like crocodiles or whales, exaggerated into myth over time.

13. The Ningen: Japan's Antarctic Cryptid

A modern addition to ocean folklore, the **Ningen** is a mysterious cryptid said to inhabit the icy waters near Antarctica. First reported by Japanese fishermen, it has become a subject of fascination for cryptid enthusiasts worldwide.

- **Description:**
  The Ningen is described as a **whale-sized creature**, pale and human-like in appearance, with long limbs or

fins. Some reports suggest it has a **human-like face**, while others describe it as more fish-like.

- **Theories:**

  - Some believe the Ningen is an **undiscovered species**, such as a giant marine mammal adapted to cold waters.

  - Others think it's a modern myth fueled by hoaxes and blurry photos circulated online.

Fun Fact: The name "Ningen" means **human** in Japanese, emphasizing the cryptid's unsettling resemblance to us.

14. The Vanishing Islands of Legend

Tales of islands that appear and disappear have sparked wonder and fear across cultures. These islands are often portrayed as magical or otherworldly, hiding secrets or riches.

- **Hy-Brasil:**
  This mythical Irish island, said to be visible only once every **seven years**, was believed to be a land of immense wealth and advanced knowledge. Some explorers reported sighting it during transatlantic voyages, though it has never been found.

- **Phantom Islands:**
  Stories of **phantom islands**, like the **Isle of Demons** near Newfoundland, persisted on maps for centuries before being disproven. These legends may have been inspired by mirages or misidentified landmasses.

Fun Fact: Scientists suggest that **volcanic islands**, which sometimes rise briefly before sinking, could explain these legends.

15. Mo'o: The Hawaiian Dragon Spirits

In Hawaiian mythology, **Mo'o** are **dragon-like water spirits** associated with ponds, rivers, and coastal areas. They are revered as **protectors** of sacred waters and feared as **powerful guardians**.

- **Role in Mythology**:
  Mo'o are said to guard water sources, and disrespecting their domains can bring bad luck or even death. Some legends describe Mo'o transforming into beautiful women to test humans' intentions.

- **Cultural Significance**:
  Mo'o are central to Hawaiian spirituality, symbolizing both the power and fragility of water. They often feature in **chanting rituals** and **oral histories**.

Fun Fact: The word "mo'o" can also mean **lizard** in Hawaiian, linking these spirits to real reptiles.

16. The Bermuda Triangle - The Supernatural Side

Beyond the theories of science, the Bermuda Triangle has long been the subject of myths and supernatural speculation. Many believe it to be a **portal to another dimension** or the **home of Atlantis**, where strange energies affect ships and planes. Others claim the disappearances are caused by **alien abductions** or the work of mysterious **sea monsters**.

In popular culture, the Triangle has become a symbol of the **unknown**, inspiring stories of adventurers, ghost ships, and underwater civilizations. Whether you view it as myth or mystery, the Bermuda Triangle remains a captivating legend of the deep.

Conclusion: Timeless Legends, Endless Mysteries

From ancient myths to modern cryptids, the legends of the sea capture the human spirit of curiosity and wonder. As we continue to explore the ocean's depths, who knows what truths might emerge to fuel the next generation of stories? The sea, vast and enigmatic, will always remain a source of awe and inspiration.

Activity: Create a Sea Legend

Materials:

- Paper, pencils, and crayons

Instructions:

1. Imagine a mythical sea creature or phenomenon.

2. Draw your creation, describe its powers or role, and explain the legend surrounding it.

3. Share your legend with friends or family to inspire even more creativity.

Chapter 10:
Beyond the Blue
- Unsolved Mysteries of the Ocean

The ocean covers more than **70% of Earth's surface**, yet we've explored less than **5%** of it.

What lies beneath the waves remains one of humanity's greatest mysteries. From **unexplained sounds to strange structures and sightings**, the ocean is full of phenomena that puzzle scientists and spark the imagination.

In this chapter, we'll uncover some of the most intriguing **real-life ocean mysteries** and the scientific efforts to solve them. Get ready to journey into the **unexplored depths** of our planet!

The Bloop: A Mystery from the Deep

In 1997, researchers monitoring underwater sounds heard a mysterious, ultra-low-frequency noise they nicknamed the **Bloop**. The sound was detected by hydrophones thousands of miles apart and was louder than any known animal or geological event.

- **What Was the Bloop?**
    - At first, scientists speculated it could have been made by a **massive marine creature**, perhaps larger than a blue whale.
    - Later, it was suggested that the sound might have been caused by **icebergs cracking** or underwater volcanic activity.
- **Unsolved Questions:**
  Even though the iceberg theory is widely accepted, some still believe the Bloop was created by an **unknown giant creature**, adding to its allure.

Fun Fact: The Bloop's location was near the **South Pacific Ocean**, close to the area where **H.P. Lovecraft** placed the sunken city of **R'lyeh** in his fiction.

2. Underwater Anomalies: Structures and Shapes

The ocean floor is dotted with **unusual shapes and structures**, some of which are natural and others that defy explanation.

- **The Baltic Sea Anomaly:**
  Discovered in 2011, this mysterious object on the Baltic Sea floor resembles a **giant disc**, sparking speculation about it being a UFO or an ancient artifact. Some scientists believe it's simply a natural formation created by glacial activity.

- **Yonaguni Monument (Japan):**
  Off the coast of Japan lies a series of **underwater stone structures** that look like a **pyramid**. While some think it's evidence of a lost civilization, others argue it's a naturally occurring rock formation.

Fun Fact: The ocean floor often creates optical illusions, making natural formations appear man-made or alien.

3. The Mariana Trench: Earth's Deepest Mystery

The **Mariana Trench** is the deepest part of the ocean, plunging more than **36,000 feet** into darkness. Despite its extreme conditions, scientists have discovered life forms thriving there, raising fascinating questions about Earth's adaptability.

- **What's Down There?**
  - **Strange Creatures:** From translucent shrimp to "ghostly" snailfish, the trench is home to bizarre life forms adapted to high pressure and no sunlight.
  - **Hidden Ecosystems:** Scientists suspect there could be entire ecosystems in the trench that have yet to be discovered.
- **Unexplored Zones:**
  Less than **1%** of the trench has been explored. Who knows what else awaits discovery?

Fun Fact: The **pressure** at the bottom of the Mariana Trench is over **1,000 times** greater than at sea level—equivalent to having **50 jumbo jets** stacked on top of you!

4. The Unexplained Disappearances of Ships and Planes

While Chapter 9 touched on the **Bermuda Triangle**, the ocean is home to many other unexplained disappearances, each with its own set of mysteries.

- **The USS Cyclops (1918):**
  This U.S. Navy ship disappeared without a trace in the Atlantic, along with its 306 crew members. No wreckage was ever found, and its fate remains unknown.

- **Flight MH370 (2014):**
  One of the most modern ocean mysteries, Malaysia Airlines Flight MH370 vanished over the Indian Ocean. Despite extensive search efforts, only small pieces of debris have been recovered.

Fun Fact: The vastness of the ocean makes it incredibly challenging to locate missing vessels or aircraft, even with today's advanced technology.

5. Unusual Ocean Sounds: Songs of the Deep

The Bloop isn't the only mysterious sound recorded in the ocean. Scientists have captured a variety of unexplained noises, each with its own story.

- **The Julia Sound:**
  Recorded in 1999, this eerie noise resembled a **muffled wail**. It lasted for over a minute and was likely caused by an iceberg, though some speculate otherwise.

- **The Upsweep:**
  A sound of rising and falling tones, detected since the 1990s, has no definitive explanation. It's thought to originate from volcanic activity or hydrothermal vents.

Fun Fact: The ocean acts as a giant sound amplifier, allowing noises to travel thousands of miles, making their sources hard to pinpoint.

6. Ocean Vortices: Nature's Whirlpools

In certain parts of the ocean, powerful whirlpools or rotating water currents, known as **ocean vortices**, occur. While most are natural, some are unusually intense or long-lasting.

- **The Dragon's Triangle (Japan):**
  Known as the **Devil's Sea**, this region near Japan is infamous for mysterious disappearances and strange whirlpools. Some claim it's the Pacific counterpart to the Bermuda Triangle.

- **Mysterious Ocean Eddies:**
  Massive eddies, or circular currents, can trap debris, marine life, and even small vessels. These eddies are often poorly understood due to their remote locations.

Fun Fact: The largest recorded whirlpool in history, known as the **Maelstrom**, occurs near Norway and was so famous it inspired stories by Edgar Allan Poe.

7. Uncharted Regions: The Abyssal Plain

The **abyssal plain** is one of the flattest and least explored regions of the Earth, covering vast stretches of the ocean floor.

- **Why It's Mysterious:**
  - These plains are covered with **layers of sediment**, hiding geological and biological secrets.
  - **Shipwrecks** and **sunken treasures** may be buried beneath the sand.

- **Possible Discoveries:**
  Scientists believe that future explorations of the abyssal plain could reveal new species, underwater volcanoes, or even resources like rare metals.

Fun Fact: Despite being considered "flat," abyssal plains feature dramatic underwater mountains known as **seamounts**, which can rise thousands of feet.

8. The Giant Squid: From Myth to Reality

Once thought to be the stuff of sailors' tales, the **giant squid** is a real creature, though it remains one of the ocean's most elusive animals.

- **The Mystery:**
    - Giant squid sightings were often dismissed as myths until scientists captured the first live footage in **2004**.
    - Despite this, much of their behavior, including how they hunt, remains unknown.

- **Fun Fact:** Giant squids are thought to grow up to **43 feet long**, with eyes the size of dinner plates—the largest in the animal kingdom!

Conclusion: The Ocean Awaits Discovery

The mysteries of the ocean remind us how much of our planet remains unexplored. From unexplained sounds and strange creatures to uncharted depths and ancient anomalies, the ocean is a place of endless curiosity and wonder. As technology improves, we may one day uncover the answers to these puzzles—or discover entirely new ones.

Activity: Solve an Ocean Mystery

Ethereal Ray

Materials:

- Paper, markers, or a notebook

Instructions:

1. Choose one of the mysteries from this chapter.

2. Write or draw your theory about what could explain it. Is it a natural phenomenon, an undiscovered creature, or something else entirely?

3. Share your ideas with friends or family and discuss how you might investigate these mysteries further.

Ocean Quiz

1. What covers more than 70% of the Earth's surface?

a) Forests
b) Oceans
c) Deserts
d) Mountains

Reference Chapter: 1 (Welcome to the Ocean)

2. Which ocean zone is closest to the surface and receives the most sunlight?

a) Abyssal Zone
b) Twilight Zone
c) Sunlight Zone
d) Midnight Zone

Reference Chapter: 2 (The Ocean Zones)

3. What phenomenon causes the glowing light produced by certain sea creatures in the deep ocean?

a) Phosphorescence
b) Sunlight reflection
c) Bioluminescence
d) Fireflies

Reference Chapter: 3 (Ocean Creatures and Phenomena)

Ethereal Ray

4. Which of the following is a sea creature that has been mistakenly identified as a mermaid?

a) Manatee
b) Shark
c) Whale
d) Dolphin

Reference Chapter: 3 (Ocean Creatures and Phenomena)

5. What is the name of the mysterious sea monster that can sink entire ships in Scandinavian folklore?

a) Leviathan
b) Kraken
c) Siren
d) Ningen

Reference Chapter: 9 (Legends of the Sea - Mythology and Folklore)

6. What is the name of the lost city that is said to have sunk beneath the sea?

a) Lemuria
b) Atlantis
c) El Dorado
d) Utopia

Reference Chapter: 9 (Legends of the Sea - Mythology and Folklore)

7. Which of the following creatures is described as a giant sea serpent in many cultures?

a) Kraken
b) Sea Dragon
c) Ningen
d) Leviathan

Reference Chapter: 9 (Legends of the Sea - Mythology and Folklore)

8. What deep-sea phenomenon is known for producing the Bloop, a mysterious underwater sound?

a) Rogue waves
b) Iceberg movements
c) Underwater volcanic activity
d) Whale songs

Reference Chapter: 10 (Ocean Mysteries and Phenomena)

9. Which region is known for the mysterious disappearances of ships and planes, often associated with the Bermuda Triangle?

a) Dragon's Triangle
b) Hy-Brasil
c) Devil's Sea
d) The Lost Island

Reference Chapter: 10 (Ocean Mysteries and Phenomena)

10. What is the name of the deep-sea trench considered the deepest part of the world's oceans?

a) Great Barrier Reef
b) Mariana Trench

c) Indian Ocean Trench
d) Atlantic Rift

Reference Chapter: 10 (Ocean Mysteries and Phenomena)

11. What ocean phenomenon is said to be the result of a massive whirlpool that can swallow ships whole?

a) Maelstrom
b) Tidal wave
c) Tsunami
d) Whirlpool of Doom

Reference Chapter: 10 (Ocean Mysteries and Phenomena)

12. The concept of a giant sea serpent in Norse mythology, encircling the world, is represented by which creature?

a) Hydra
b) Kraken
c) Jörmungandr
d) Leviathan

Reference Chapter: 9 (Legends of the Sea - Mythology and Folklore)

13. What is the primary reason mermaids are often depicted as luring sailors to their deaths in folklore?

a) They want revenge on humans
b) They are protecting the seas
c) They are attempting to gather souls
d) They are simply curious about humans

Reference Chapter: 9 (Legends of the Sea - Mythology and Folklore)

14. The giant squid is thought to be the inspiration for which sea monster legend?

a) Kraken
b) Makara
c) Leviathan
d) Ningen

Reference Chapter: 10 (Ocean Mysteries and Phenomena)

15. What is the name of the phenomenon that creates glowing lights on the masts of ships during thunderstorms, believed to be a sign of divine protection by sailors?

a) Aurora Borealis
b) St. Elmo's Fire
c) Moon Halo
d) Lightning Bugs

Reference Chapter: 9 (Legends of the Sea - Mythology and Folklore)

16. What deep ocean phenomenon is known for its extreme pressure and dark, cold waters, home to the most mysterious and alien-looking creatures?

a) Twilight Zone
b) Sunlight Zone
c) Midnight Zone
d) Abyssal Zone

Reference Chapter: 2 (The Ocean Zones)

Ethereal Ray

17. What legendary underwater creature is said to have the ability to change shape and is part of both Scottish and Irish folklore?

a) Selkies
b) Mermaids
c) Leviathan
d) Sirens

Reference Chapter: 9 (Legends of the Sea - Mythology and Folklore)

18. Which of the following is a real-life phenomenon that is thought to cause rogue waves or unexpected large waves at sea?

a) Ocean currents meeting
b) Underwater volcanoes
c) The moon's gravitational pull
d) Meteorological storms

Reference Chapter: 10 (Ocean Mysteries and Phenomena)

19. What is the primary factor that influences the movement of ocean currents across the globe?

a) Wind
b) The Moon's gravitational pull
c) Water temperature
d) Earth's rotation

Reference Chapter: 2 (The Ocean Zones)

20. The Great Conveyor Belt, which is responsible for moving warm and cold water across the globe, is part of what process?

a) Thermohaline circulation
b) Oceanic upwelling
c) Sea-floor spreading
d) Oceanic evaporation

Reference Chapter: 2 (The Ocean Zones)

Answers:

1. b) Oceans

2. c) Sunlight Zone

3. c) Bioluminescence

4. a) Manatee

5. b) Kraken

6. b) Atlantis

7. b) Sea Dragon

8. c) Underwater volcanic activity

9. c) Devil's Sea

10. b) Mariana Trench

11. a) Maelstrom

12. c) Jörmungandr

13. a) They want revenge on humans

14. a) Kraken

15. b) St. Elmo's Fire

16. c) Midnight Zone

17. a) Selkies

18. a) Ocean currents meeting

19. a) Wind

20. a) Thermohaline circulation

Glossary of Ocean Terms

Abyssal Zone
The part of the ocean floor that lies between **13,000 feet** and **36,000 feet** deep. This zone is the most extreme environment, with high pressure, low temperatures, and no light. It is home to strange and unique species that are adapted to survive in these harsh conditions.

Bioluminescence
The production and emission of light by living organisms. Many marine creatures, such as jellyfish, squid, and certain types of plankton, use bioluminescence to attract mates, lure prey, or deter predators.

Conveyor Belt (Thermohaline Circulation)
A global system of ocean currents driven by temperature and salinity differences in seawater. It helps regulate Earth's climate by moving warm water from the equator to the poles and cold water from the poles back to the equator.

Currents
Continuous, directed movements of seawater generated by forces such as wind, temperature differences, and the rotation of the Earth. Surface currents are influenced by wind, while deep-water currents are driven by temperature and salinity differences.

Depth Zones
These are divisions of the ocean based on depth, temperature, and the amount of sunlight they receive. The **Sunlight Zone** is closest to the surface and supports life that relies on sunlight. Below that are the **Twilight Zone**, **Midnight Zone**, and the extreme **Abyssal Zone**.

Ethereal Ray

Hydrothermal Vents
Underwater fissures or openings in the Earth's crust that emit hot, mineral-rich water. These vents support unique ecosystems of organisms that rely on chemosynthesis (using chemicals from the vents for energy) rather than sunlight.

Kraken
A legendary sea monster from Scandinavian folklore, often depicted as a giant squid or octopus capable of dragging entire ships and their crews into the depths of the ocean.

Leviathan
A giant sea creature mentioned in various ancient mythologies, particularly in the Bible, where it is described as a symbol of chaos and divine power. The Leviathan is often depicted as a massive serpent or dragon.

Lemuria
A mythical sunken continent, believed to have existed in the **Pacific or Indian Ocean**. The legend says it was home to an advanced civilization that disappeared when the continent was submerged under the sea.

Manatee
A large, slow-moving aquatic mammal, sometimes mistaken for a mermaid in folklore. Manatees are herbivores and are found in coastal waters and rivers, primarily in warm tropical regions.

Mermaid
A mythical creature with the upper body of a woman and the lower body of a fish. Legends of mermaids exist in many cultures, often portraying them as either benevolent protectors or dangerous seductresses.

Ningyo
In Japanese folklore, the **ningyo** is a mermaid-like creature. It is often depicted as having the body of a fish and the face of a human. Eating the flesh of a ningyo is said to grant immortality, though it also brings misfortune.

Rogue Waves
Unusually large waves that are much higher than the surrounding waves and appear unexpectedly. Rogue waves can be extremely dangerous and often cause damage to ships and offshore platforms.

Scylla
In Greek mythology, Scylla is a monstrous sea creature with multiple heads that devours sailors who come too close. She is one of the two dangers in the **Strait of Messina**, alongside **Charybdis**.

Sirens
In Greek mythology, Sirens are sea creatures who lure sailors to their doom with their enchanting songs. Unlike mermaids, Sirens were often depicted as half-bird, half-woman creatures.

Sunlight Zone
The top layer of the ocean, extending to about **650 feet**. This zone receives the most sunlight and is where most ocean life is found, including many fish, plants, and other creatures that depend on photosynthesis.

Tidal Wave
A large ocean wave, typically caused by undersea earthquakes or volcanic activity, capable of causing significant damage along coastlines. Also known as a **tsunami**, a tidal wave is much more destructive than typical waves.

Twilight Zone
The middle layer of the ocean, where light begins to fade and is no longer sufficient for photosynthesis. This zone is home to creatures that can live with very little light and must rely on other sources of energy.

Upwelling
The process by which cold, nutrient-rich water from the deep

ocean rises to the surface. Upwelling zones are often highly productive ecosystems, supporting rich marine life.

Whirlpool
A rapidly rotating body of water, often formed by opposing currents or tides. Whirlpools can be dangerous, and some of the most famous ones, like the **Maelstrom** in Norway, are the subjects of legends.

Zooplankton
Tiny, often microscopic creatures that drift in the ocean, feeding on algae or other small organisms. Zooplankton is an essential part of the ocean food chain, serving as food for fish, whales, and other marine creatures.

Abyssal Plain
A flat, nearly featureless area of the ocean floor, covered in thick layers of sediment. Abyssal plains cover about **40% of the Earth's surface** and are typically found at depths greater than **13,000 feet**.

Marine Snow
A continuous, slow fall of organic matter, such as dead plankton and other debris, from the upper layers of the ocean to the deep ocean. This material forms the base of the food chain for many deep-sea organisms.

Tsunami
A series of large ocean waves caused by undersea earthquakes, volcanic eruptions, or landslides. Tsunamis can travel across entire ocean basins and cause massive destruction upon reaching coastlines.

Hydraulic Press
The immense pressure in the ocean, especially at great depths, can crush objects and even create underwater features like vents. Hydraulic pressure in the deep sea is essential to understanding the unique conditions in regions like the **Abyssal Zone**.

Ghost Ships
Vessels found adrift with no crew aboard. Some famous ghost ships include the **Mary Celeste** and the **Flying Dutchman**. These tales of vanished crews have sparked many theories, ranging from natural causes to supernatural influences.

The Great Barrier Reef
A massive coral reef system located off the coast of **Australia**, famous for its biodiversity. It is the largest living structure on Earth, stretching over **1,430 miles**, and it plays a crucial role in marine ecology.

The Great Ocean Conveyor Belt
A large-scale ocean circulation driven by differences in temperature and salinity. This "belt" helps to regulate global climate by redistributing heat and nutrients across the world's oceans.

Upwelling Zones
Areas in the ocean where cold, nutrient-rich water rises to the surface, often supporting large populations of marine life. **Upwelling** is important for the productivity of fisheries and is a key factor in the health of the ocean's ecosystems.

Conclusion: The Ocean's Wonders

As you've journeyed through the pages of this book, you've learned about the incredible vastness of the ocean, the wonders beneath its waves, and the mysterious phenomena that continue to fascinate us. From the sunlit surface to the mysterious depths, the ocean is a vast and dynamic world full of life, discovery, and unexplored mysteries. Whether it's the sea monsters of ancient legends or the real-life wonders of the ocean floor, the ocean will always remain an endless source of fascination.

Remember, just as the stars above us are made of stardust, the sea, too, is a place of infinite wonder, where new discoveries wait to be made. As we continue to explore and

Ethereal Ray

protect our oceans, we honor the mysteries they hold and the life they sustain. So, dive deep into this magical world and keep your curiosity alive!

Review Request

Did this book **make a splash in your imagination?** If so, Finn and I would love to hear from you! Your review on **Amazon or Goodreads** helps other young explorers **discover the wonders of the ocean** and inspires us to **keep uncovering new underwater secrets!**

Share your favorite discoveries! If you had fun reading, consider uploading a **photo or video** with your Amazon review! Whether it's a **mind-blowing sea creature**, a **favorite page**, or even a **drawing of an underwater world you imagined**, we'd love to see it!

   Leave a review by scanning the QR code below! Your insights **help us keep exploring** and might even inspire the next volume of *The Ultimate Ocean Facts Book!*

Thank you for joining this **deep-sea journey—stay curious, keep exploring, and never stop wondering about the ocean's mysteries!**

Ethereal Ray

Check my other ongoing books in the series!

Scan the QR below

Made in the USA
Monee, IL
25 June 2025

19987966R00059